Joyce
I hope you enjoy the farm!
3-3-21

THE FARM AND OTHER TRUE STORIES

Wendy West

Wendy West

Turtle House Ink

Printed in the United States of America

Cover: "The Farm" by Hannah West O'Connor (1986)

Photo: "The Farm and Other True Stories" by Christina Patoski

ISBN 978-0-9666632-2-8

Turtle Ink House

235 Black Oaks Lane
Wayzata, Minnesota 55391

The Farm
and Other True Stories

In memory of Roddy West
January 13, 1948 - May 10, 2020

For my family

"He who works with his hands is a laborer.
He who works with his hands and his head is a craftsman.
He who works with his hands and his head and his heart is an artist."*

*Note: The true origin of this quote is unknown. It has been falsely attributed to St. Francis of Assisi but cannot be found in any of his writings. It first appeared in a 1923 publication, *Boys' Workers Round Table*, and was later quoted in a 1948 book by Louis Nizer, *Between You and Me*.

Contents

THE FARM

Prologue

It was 1969. Roddy and I had been together for less than a year. We were against the Viet Nam War and regularly took part in demonstrations. On December 1, the lottery for the draft was announced. Roddy's number was three hundred eighteen. With such a high number, it was unlikely that he would be drafted.

"Now we can spend the summer traveling," Roddy said as we were driving into the city after having dinner with his parents.

"My dad won't let me go traveling with just you," I said. "We aren't married."

"Well, I wouldn't mind marrying you." That is how he proposed to me.

We were married in August of 1970. I finished college, and Roddy was still taking a few classes that had been canceled when the university closed because of the strike against the war. We both had part-time jobs. In the summer of 1971, we found full-time jobs in the federally funded alternative school project called the Southeast Alternatives. There were five school choices that parents could choose from based on their own educational philosophy. We worked at the Free School, the most liberal of the choices, where parents and teachers were the governing board and made all of the decisions about school policies and the curriculum. I was one of three primary teachers, teaching kindergarten through third grade; and Roddy was one of three middle school teachers teaching fourth through eighth grade.

Teaching gave us free time in the winter, spring, and summer. We would have time to pursue new interests. Roddy read a review of Helen and Scott Nearing's book *Living the Good Life* in the *Whole Earth Catalog*. The subtitle of the Nearing's book was *How to Live Sanely and Simply in A Troubled World*. That resonated with us. The Nearings expressed a desire to make a living, follow their political beliefs, and have time for leisure and other pursuits of interest. That appealed to us too.

We were not as idealistic or philosophical as the Nearings. We were young, in love, had just finished college, and wanted to do something different. We had both grown up in affluent midwestern homes, liked the Midwest, and had given up on the idea of traveling the world. We were enjoying our lives teaching. We had made a cedar strip canoe and were happy to stay put. We thought it would be fun to have a place to go for weekends and school breaks. We began to look for land in Minnesota and Wisconsin, hoping to find something two to three hours from Minneapolis. We looked at lots of farm houses, some on busy roads, some on lonely gravel roads. When we found the abandoned farm in Timberland, we marveled at the beauty: rolling hills, open pastures, ravines, a forest, and a small lake. We knew this was it.

The Farm

Roddy and I bought the farm in the spring of 1973. It was 120 acres, two hours northeast of Minneapolis in an area called Timberland. The homestead was covered with oaks, maples, and a few virgin white pines marking the corners of the three parcels. The rest was fields and pasture. The ravine drained into Places Lake, a spring fed watering hole for the cows that grazed in the pastures.

A red barn filled with hay and a small milk house connected to it sat near the fields. We could smell freshly mown hay and ripe cow pies. The Odin farmhouse had been abandoned for almost twenty years and was collapsing in on itself. The rain had soaked through the wood and what was left of the insulation gave off a moldy rotten smell. Pots were scattered in what was once the kitchen and broken dishes surrounded the collapsed table. Hollyhocks and thistles poked through the floorboards.

We spent hours walking the land to decide where we would build our "shabin," half shack, half cabin. We decided to pick a spot near the ravine where we could see the lake and the open pasture and fields to the west. We would have beautiful sunrises as well as sunsets.

We were teaching and had the summer off. It all started with building the outhouse, using old wood from the farmhouse. We slept in a three-room canvas tent, nicknamed "The Hilton." We had to fence ourselves in from the curious cows grazing in the surrounding pasture. We learned the hard way. One night, we went into town to see the movie *Klute* with Jane Fonda. We returned after dark and walked up the road with our flashlights. The lean-to tarp of our kitchen was destroyed. Our pots and pans were stomped and pooped on and the cows had even walked through our folding canvas chairs.

Once the outhouse was built, the next step was to collect rocks from the numerous piles in the hayfields and in our favorite spot, the back forty. We had read Helen and Scott Nearing's book, *Living the Good Life* and decided to use their method for building stonewalls. We would climb into our rusty red '54 Chevy and head out to the rock piles and fill the bed as full as we could, always thinking about the precious axle.

Again and again, we carried the rocks back and arranged them by size. The rocks were beautiful colors, and we always had favorites. Next were our trips to Mrs. Bly's gravel pit to get gravel and sand for making our cement. Mrs. Bly was a widow who lived in a beautiful stone house next to a wonderful sounding brook. She was an avid reader and seemed interested in our pursuit. She charged us twenty-five cents a load. Collecting and sorting rock and shoveling sand was grueling work, but we loved it. I felt so strong.

Our cement mixer came from Ernie Lauritsen, the farmer who rented our fields to raise oats, hay, and corn and our pastures to feed his young stock. Ernie was rugged and gentle. He had lived on this land in Timberland all of his fifty years. His day started at five in the morning with his chores: cleaning the stalls, washing the milking machine, and getting silage ready for the cows. He had to be ready to deal with the unexpected. A cow that developed milk fever, a flat on the tractor, or a gash in his arm from barbed wire as he stretched a piece of broken fence, nothing seemed to faze him; he was always smiling and whistling.

Ernie was married to Dorothy, a kind, wonderful cook and a hard worker. She taught me how to make bread and butter pickles and corn relish. Dorothy was like a mother and treated me as an adult. I wasn't sure I wanted to be a farm wife, but I craved our friendship. She was curious that I had gone to college and was a teacher, things she knew nothing about. I loved sitting at her kitchen table watching her make dinner, the noon meal. There were always potatoes, some kind of meat with gravy, not many vegetables unless they were fresh from her garden, and desserts galore. Dorothy had a dessert cookbook called *The Little Red Book* published by the Ladies' Aid of her church. Our favorite was Top Me Twice Cake. It was a pineapple upside down cake that you poked holes in when it came out of the oven and poured a mixture of powdered sugar, butter and milk into the holes.

We started building our house. First we built forms for the walls with used lumber. I spent hours pulling and straightening nails and stored them in coffee cans. The first stonewall for the "shabin" was fourteen-feet long, and we arranged the rocks in sections about six-feet long and three-feet high. I was in charge of mixing the mud: a bag of cement, eight shovels of sand and just enough water to make a thick batter. I loved the smell and was tempted to taste it many times.

We listened to the Watergate hearings, as well as reports about the protests at Wounded Knee on a large battery operated radio my father had given us. Public radio was our main connection to the rest of the world from our idyllic perch above Places Lake. When Nixon resigned on August 9, 1974, we quit working and jumped in our car and drove back to Minneapolis to celebrate.

When the stonewall was finished, we moved two walls from the old farmhouse up the hill on the back of a hay wagon. The fourth was built with lumber from the local junk dealer. Next came a screened porch, where we loved to sleep especially on nights when thunderstorms rolled through the valley.

When school started, we moved back to our apartment in the city. We returned to the "shabin," on weekends, hiked through the woods, and got to know many of our neighbors. We purchased a wood stove from the junk dealer and cut wood in the ravine. We spent Christmas break there, inviting family and friends to sleep in the loft and cross-country ski.

* * *

I got pregnant in February, and we thought about quitting our jobs and living on the land. June came and we quit our jobs. Our plan was to move the old farmhouse from a neighbor's yard and attach it to our "shabin." The adventure began. We had no idea what directions our lives would take. All we knew was that we were leaving the city and wanted to be self-sufficient: to live off the land, enjoy its beauty, and be far away from the city. We had no idea how hard it would be or how we would fit into the community. We put in a well, brought in electricity, and added a septic tank.

The land cost $15,000, which we borrowed from my father and paid him back within a year. We weren't really thinking that this would be our life forever. We were just enjoying the experience. It was so different from anything we had ever done. We liked being in charge, planning, and figuring out how to do all the tasks.

My grandmother had left me some General Electric stock, so we used that to pay for putting in electricity, our well, and septic system. At first our parents thought it was a fun idea that we would have a place to spend summers, but once we quit our jobs and moved their feelings changed.

My father asked Roddy if he was taking me to Tobacco Road and thought this kind of life was irresponsible because we were having a baby. My mother loved to come and help with building projects. Roddy's parents occasionally stopped by on their way to visit friends, but they were so uncomfortable they never sat down.

Every weekend my brother Edmund would drive four hours to spend the weekend working with us on the farm. We were trying to get the house ready to move in to before winter and before our baby arrived. Edmund lived in a tent we called the little blue pup. He left it closed up all day and then every night before he crawled in, he would fog it with Raid. We wondered how many brain cells he was killing.

By September, we had moved the house from Sidney Brecke's yard. The house was the original Timberland post office and then the Brecke homestead. Sidney and Betty had built a split-level house in the sixties so the L-shaped farmhouse was empty. The Breckes were happy to get it out of their yard. They gave it to us for free. We just had to move it. With the help of Eldo Bengston, who charged us $500, we cut it in half, jacked it up

on a large hay wagon, and rolled it a mile and three quarters to the foundation we had built with fieldstone.

We bartered for most of our building supplies from Tony Orlicki, the local junk dealer on County Road H. Roddy helped Tony unload his truck on Thursdays so we exchanged labor for many of our supplies and furnishings like our cast-iron claw foot bathtub, laundry tubs, and our lovely metal lawn chairs.

On Saturday nights the three of us would get all cleaned up and go to the Barronett Supper Club to eat ribs and au gratin potatoes. We would watch all of the farmers dance polkas and schottisches. We were too timid to dance.

As winter approached, we had moved into the house, and Hannah was born, Edmund stopped coming every weekend. He had a girlfriend, Kathy, and he spent all of his time with her. The next thing we knew they were getting married and planning a June wedding. My father was furious that Edmund hadn't told the truth about Kathy being pregnant. This was a bad start to their relationship with my father and to their marriage.

Sidney Brecke had given Roddy a .22 rifle and told him how he could shoot a deer at dusk in our fields.

"They've been eating your oats all summer. Just one pop between the eyes, but you need to be careful because the game warden comes around," Sidney told Roddy.

I definitely did not want Roddy going hunting during deer season. We had heard horror stories of guys sitting in bars and then going out in the woods and shooting at anything that moved.

For two weeks Roddy went down to our main field and sat on a rock pile watching the deer until dark. I felt he was avoiding me. We had checked a book out of the Cumberland Library that explained how to field dress a deer and how to cut up and package the meat.

One night he came running into the house, got a flashlight, and took the book down to the field. He was panting.

"I dragged it to the rock pile with my belt," he told me.

He wanted to get the hide off as quickly as possible to cool the meat. He buried the entrails and head and brought the deer up to the house in the back of our pickup. He used an electric saw to cut it in pieces small enough to put in our refrigerator in plastic bags. I felt as if we were criminals and that the game warden would be knocking on our door.

We followed the instructions in the book and cut the meat into steaks, chops, and stew meat. We borrowed a meat grinder from the Breckes and had packages of ground venison. It looked so professional wrapped in butcher paper, marked and dated. We discovered we could not eat it more than three times a week, or we would have terrible gas. Borscht soup was one of our favorites as well as stew with our carrots, potatoes, onions, and herbs.

Once winter set in our friends from the city stopped coming to visit, so the days were long, cold, and darkness came early. We spent a lot of time playing with and just watching Hannah. I describe this period as our baby-mush-brain time. I rarely thought about anything but Hannah. She was beautiful and such a happy baby.

Hannah had an outie belly button. It was so large we could see movement. My parents stopped by one Sunday on their way home from their cabin, and my father took one look at her stomach.

"God, you have to bind her stomach. Get a fifty-cent piece and press it so her intestines go in. That's a hernia," he told me.

I was in tears. We were thirteen miles from the hospital, and it was Sunday. What was going to happen?

The next afternoon my father called and said he had spoken to one of his colleagues who was a pediatrician, and there was no need to worry or to bind Hannah.

"In fact, if you were living in certain parts of Africa, it would be considered a sign of beauty," he said.

I joined Book–Of-the-Month Club and checked out stacks of books at the library. My bible was *Dr. Spock's Baby and Child Care* that I kept by the rocking chair and read as I nursed Hannah. Doris Lessing was a favorite and I read *Helter Skelter* about Charles Manson. We didn't have television but listened to music and news and public radio.

We loved to put Hannah inside Roddy's jacket in the Bjorn and go cross-country skiing all over our hundred and twenty acres.

* * *

In the spring, we decided we would collect maple sap from the trees in the ravine and make maple syrup. We went to the Anderson Sugar Bush near Stillwater, and bought tubing, spigots, and buckets. The staff graciously told us where and how to tap the trees.

We bought a large rectangle, flat metal pan from Tony Orlicki and set up our operation at the foot of the ravine. We moved piles of wood and built a rack with fence posts to hold the pan so we could put fire under it. It takes forty gallons of sap to make one gallon of syrup. We had absolutely no idea how much we would get, but we were excited about trying to collect maple syrup.

We would bundle Hannah up and put her in a small wagon where she slept while we got things set up. We couldn't control the soot and ash from getting into the boiling sap. We read about straining it through cheesecloth, but the soot darkens the sap so the quality is not as good. We actually got about two quarts, and it tasted delicious. We planned that we would do more as long as the weather cooperated. Unfortunately the temperatures started getting too warm and the sap stopped running. We would do more research and try it again the next year.

* * *

We woke up early on Sunday morning, April 5, 1975. The sky was a brilliant blue, and we were sure that the temperature would raise enough to melt the snow that made our driveway impassable. Roddy started a wood fire in the furnace, and I began packing baby clothes to take to

Minneapolis to exchange with Roddy's two sisters who had children close to Hannah's age. We were excited to see the cousins together. Hannah was four months old and had just learned to roll from front to back and then from back to front. She was pleased with herself, and we were definitely proud parents.

I noticed smoke coming from the hay bale that was covering our wood chute into the basement. I told Roddy and he went outside and found that the shingles near our chimney were burning.

" Get on the phone and call Dorothy to see if they have a large fire extinguisher for the barn," he said.

I called Dorothy, our closest neighbor, who said they did not. We hooked up a hose and began filling buckets of water. I put Hannah in her snowsuit and set her outside on a large pillow. She fell right to sleep. Our cat, Frances, stayed next to her.

The access to the upstairs had been cut off when we had cut the shorter portion of the house to move it, and the only way to get into the second floor was by using a ladder.

Roddy put the ladder up next to the door and carried up the diaper pail filled with water. When he opened the door, flames shot across the opening engulfing all of the space. He barely made it down before the flames leaped out of the door.

Just then we saw nine farmers walking up our driveway. Dorothy had called them on the party line. They started carrying everything out of the house. By now the whole roof was burning.

"Please don't go inside! There is nothing we need. I really don't want anyone getting hurt," I said. I was hysterical.

I watched as the men carried out my wedding china and set it on the tailgate of our pick up. Three men lifted our freezer full of meat and vegetables out of the basement. It was getting so hot that the windows were popping, glass falling everywhere. The men decided they needed to move the pick-up farther away from the house. All of the dishes crashed to the ground. Only one bowl, a dinner plate, and a salad plate survived.

The house was completely gone in forty-five minutes. We were in a state of shock.

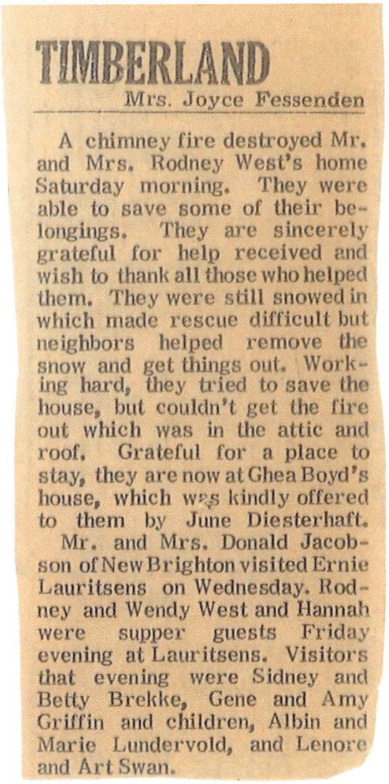

TIMBERLAND

Mrs. Joyce Fessenden

A chimney fire destroyed Mr. and Mrs. Rodney West's home Saturday morning. They were able to save some of their belongings. They are sincerely grateful for help received and wish to thank all those who helped them. They were still snowed in which made rescue difficult but neighbors helped remove the snow and get things out. Working hard, they tried to save the house, but couldn't get the fire out which was in the attic and roof. Grateful for a place to stay, they are now at Ghea Boyd's house, which was kindly offered to them by June Diesterhaft.

Mr. and Mrs. Donald Jacobson of New Brighton visited Ernie Lauritsens on Wednesday. Rodney and Wendy West and Hannah were supper guests Friday evening at Lauritsens. Visitors that evening were Sidney and Betty Brekke, Gene and Amy Griffin and children, Albin and Marie Lundervold, and Lenore and Art Swan.

All of a sudden Sonny Boyd, who I knew only by name, came up to us and said, "My mother-in-law passed away a few months ago, and her house is empty. You are welcome to stay there until you decide what you want to do."

We took our possessions in truckloads over to the little square blue house. It had a barn and a chicken coop and backed up to beautiful state land with lots of trails. It wasn't until we had unloaded the third load that we realized there was an outhouse and no indoor plumbing.

"It will be just like camping. We'll make it work," Roddy said in a voice so hoarse from smoke and heat, I could barely hear him.

We stayed in the little blue house for six months. We felt such gratitude and closeness with all of our neighbors. There was no way we wanted to leave and go back to the city.

I called my mother from Dorothy's. "Mom, our house has burned to the ground. We're okay and one of our neighbors has offered us his mother-in-law's house to live in." I waited for her answer.

"Oh dear. That's too bad," she said in a voice that told me she thought I was teasing. Our family had a history of making phone calls that were totally scary just to hear the other person's reaction.

"It's true, Mom. We are still going to Minneapolis to see Roddy's sisters, and we need to talk with the insurance company. We lost all of our books and records and everything that was stored upstairs in the house."

"Are you serious?" she said in a stern voice.

"Yes," my voice was cracking. It had finally hit me that we had lost our house and so many things. I started to cry and just sat and sobbed.

"I'm so sorry, Wendy. I thought you were teasing. Let us know how we can help. Call me when you get to Minneapolis."

We went back over to our house to check the site to see if there was anything we had missed that was salvageable. I looked where our bed had been and for my grandmother's locket that had been on my nightstand. I kicked the ashes, the stench was bitter, but it wasn't there. Had it melted because of the heat? Within the locket was a four-leaf clover that my grandmother had since she was ten.

* * *

We found a "new house" in the Missner's cornfield. It was a two-story rectangle with windows, a roof, and hardwood floors that were original. The house had been used to store corn and then lumber. Large holes were in the walls where rats had chewed their way in. There were three rooms upstairs, one would be the bathroom and two would be bedrooms. The house had lots of windows and good light. It was perfect. We paid five hundred dollars and hired Eldo Bengston again. Sonny Widenberg used his bulldozer and dump truck to help us clean up the site where our first house had been. We had to enlarge the foundation by two feet on either end, but we decided to use cement blocks to get our project finished quickly.

We spent the summer insulating and putting in electricity and plumbing. Our days were filled with building and taking care of our garden. We moved in September, had new birch cabinets, a dishwasher, and I was pregnant again. Hannah was such an easy baby; we thought it would be fun to have another child close in age.

* * *

Roddy began spending more time helping Ernie with chores. Dorothy had stage four breast cancer and was unable to work in the barn. Roddy would be up at 5:00 a.m. to help Ernie. In exchange for his work, Ernie gave him bull calves. The calves would have just been thrown in the field to die. They were of no use to the dairy herd. Ernie also gave Roddy three heifers that had given birth too early to be satisfactory for his herd. Ernie liked his stock to be really large, so he bred them a year later than many of the local farmers so that the cows would be strong and produce more milk.

Roddy raised veal calves. We named the three heifers Fern, Rose, and Daisy. The heifers would nurse the calves until they were 225 pounds. If they went over that, they were considered baby beef and the price would drop a third. Hannah rode with Roddy when he drove to the stockyard. She loved to talk to the calves that were riding in the back of the pickup.

I contributed to our family income by making Raggedy Ann dolls in all sizes and doorstop chickens that I sold to local craft stores and in Minneapolis. I started dyeing infant and toddler t-shirts bright colors in an old wringer washing machine and silk-screening animals on them. My favorites were cows, sheep, and goats. Occasionally I made special orders. I drew a Paul Bunyan for Hello Minnesota, a new store in Minneapolis. They weren't huge sellers, but I loved telling people my shirts were in this new trendy store.

There were two other young women in the area, Madeline and Lynda, who had toddlers, and we became friends. Both of them had grown up in different parts of the country and had ended up in Timberland because of their husbands' work. Madeline was from Indiana and her husband was a carpenter who liked to hunt and fish. Linda had lived in Dallas and met Les, her husband, when she visited her parents in Timberland. He was dairy farming with his father.

We avoided conversations about religion and politics. We talked about food, parenting, and local gossip. They both belonged to the local Lutheran Church and a group called the Secret Pals Club. We would get together once a week for coffee and let the kids play. If one of them canceled, I would be so sad. I looked forward to a little adult conversation. Sometimes I felt very lonely.

* * *

I went to my doctor's appointment two days before my due date.

"This baby is really small and probably won't be ready for a couple of weeks, which is good because I'm going to Phoenix tomorrow for a week," the doctor told me.

I decided I was going to be as active as I could and get this baby to come out. I painted the screen porch and washed all the bedding so we could sleep on the porch.

Two days later my contractions started in the afternoon. Around midnight they were getting more intense and were three minutes apart. I woke Roddy and said I thought we should leave so that we could drop Hannah at the Lauritsen's. As we pulled into Ernie's driveway, all of the cows were out of the barn. They were around the house, in the garden, and some were in the shed.

Roddy jumped out of the car and said, "I have to help Ernie get them back in the barn."

I took Hannah who was sound asleep into the house where she snuggled with Dorothy, and I went back to the car and waited.

My contractions were getting stronger, and I was getting anxious that we might not make it the thirteen miles to the hospital. Finally all the cows were in and we were off.

"I just need to tell you something," Roddy said. "I hate the name Orissa."

That was the name we had picked if the baby was a girl. I loved the sound of Rissa West, and it had been next to Hannah's name embroidered on the antique friendship quilt we had found in Grandma West's attic.

There was a two-bed labor room in the Cumberland Hospital. We settled in. We had brought our music and asked for ice chips. I was a champion at breathing techniques. At about 1:30 a.m., Brittany was admitted and put in the other bed. She was sixteen and all alone. Every time she had a contraction, she would scream and jump out of bed.

The nurses were not very nice to her. They made faces and rolled their eyes. I felt sorry for her, so I decided to teach her some breathing techniques to calm her down. At 4:00 a.m., things began to progress for me, and I was wheeled into the delivery room. The doctor appeared and wasn't excited about Roddy coming into the delivery room.

Leah was born at 4:45 a.m. She was six pounds and twenty-one inches long. We were so excited and were finally taken to one of the two maternity rooms. We slept for a while and then were awakened by Brittany saying, "I want to be with Wendy."

The nurses brought her and her baby Amanda into my room. About 9:00 a.m., Brittany's relatives started arriving to see her and her baby, no mention of any father. They were a rough looking bunch, and it was uncomfortable to be in the room with them. I couldn't have my baby in the room while they were there.

"Hospital policy," the nurse said. I waited and waited for them to leave and then another rowdy group would show up. Finally around 4:00 p.m., I went out into the hall and burst into tears.

"I have been waiting all afternoon to be with my baby," I told the new nurse, who had come on duty.

She put her arm around me and said, "We'll just move you to the other room, and you can be with your baby as long as you like."

After that I just wanted to get home and be with Hannah and see her reaction to her new baby sister.

* * *

We decided to raise chickens to eat, not for their eggs. Roddy built a chicken coop where the chickens could be safe and rest at night.

From the co-op in Cumberlande, we bought thirty day-old chicks: fifteen Rhode Island Reds and fifteen Longhorns. Hannah loved chasing the chickens and they grew quickly. It takes a chicken about ten weeks to reach maturity.

When it came time to slaughter the chickens, we again consulted a book from the library and decided we would do it in one day because if there were blood around it would attract predators like fox and mink. We set up our processing station in the driveway. Roddy cut their throats over a garbage can so that all of the blood would run into the can. I had a large canning kettle with extremely hot water that I dunked them into and pulled off all of their feathers. The smell was horrendous. I had to stop periodically to breast feed Leah.

When Hannah woke up from her nap and came outside, she kept saying, "Chick, chick," and kissing the dead chickens' feet. I thought I had scarred her for life.

* * *

We worked hard every day, and all the work was just to keep our existence going. We had no grand plans of being farmers, doing truck farming or community sustainable agriculture. We just raised food for ourselves.

After Leah was born, the work seemed harder and less satisfying to me. The summer and fall were full and beautiful, and we loved being on the farm. The winter began to take its toll on me, and I really wondered if I could do another season without going nuts. We had been living on the farm for three years now.

Roddy tried different jobs. He worked as a carpenter for Bob Albee, but was laid off because there wasn't enough work for a crew of three. He found a job with the Comprehensive Employment and Training Act program and organized a crew of unskilled men to do weatherization projects for qualified residents in Burnett County. Some days were comical. There was one guy who called himself Scorpio Man. He left a suggestive note on the pillow of a woman the team was doing work for. The woman freaked out. Roddy had to deal with her and reprimand Scorpio Man.

He decided that he would go back to school and get a teaching certificate through Mankato State University. They gave him credit for the years he had taught at the Free School. The biggest requirement was a quarter of full-time student teaching. He taught with Dan Eiller, who had taken a sabbatical and had worked at the Free School with Roddy. Dan taught social studies in the Roseville High School, and Roddy wanted to concentrate on social studies. It was perfect.

We would pack the girls up and spend the week in Minneapolis, living in Roddy's parents' attic. The girls played with their cousins, and I sewed and took classes in silk screening. On Fridays, we would go back to the farm and do our chores. We planted another huge garden.

I began to wonder why we were living on the farm. Things in the community began to trouble me. Ernie and Dorothy's oldest daughter was very involved in the little church at the top of the hill. She participated in Bible study, taught Sunday school, and at the urging of her pastor went around to the neighbors telling people to get rid of their Ouija boards. They were the devil's tools. I loved Ouija boards. I began to feel we didn't fit in. Many of the girls aspired to be hairdressers and wanted get married right after high school. Hannah was about two-and-a-half and Leah was a year. I wanted them to be independent women, to be able to follow their dreams, and to be affirmed for their views. I didn't think they would fit in.

We weren't religious, and I didn't get my hair done.

Other things began to bother me about this bucolic setting. There was Robert Winthrop, the wife beater and child molester who lived two miles behind us. He had molested his stepdaughter, beat his stepson with a rubber hose from the barn, and smacked his wife who pretended she had bumped her head on her car door. Everyone in the community knew this and nothing was done about it. He had three little girls. They would eventually ride the school bus with Hannah and Leah. I couldn't let my girls go and play at his house.

We decided that we would move back to the city. Roddy started looking for a teaching job, and we searched for a house near Powderhorn Park. We didn't tell anyone in Timberland about our plans. We realized being subsistence farmers was not the life we wanted to live. I needed more interaction and wanted to be around people who thought more like we did.

We would keep the farm, have our gardens, and get firewood to heat our house in Minneapolis. We would spend as much time as possible there in the summer and close it for the winter.

When we returned to the city, Roddy worked for a small contractor remodeling houses in south Minneapolis. I taught part time at an alternative elementary school. Hannah and Leah went to preschool. We liked being involved with our community and having the girls be able to play with their cousins and spend more time with their grandparents.

* * *

Looking back some forty plus years, I think we wanted to change our lives and live differently than we were raised. I didn't realize how lonely and hard it would be. Once we moved back to Minneapolis, I enjoyed being at the farm more. First, because we weren't there in the winter and our stays were short. Second, we were there for specific purposes: to tend our garden and get wood to heat our house in Minneapolis. Our kids would bring friends, and we were able to spend more time having fun, hiking, going swimming, and sitting by the campfire.

We loved living off the land, being so conscious of the weather, and working our bodies to the point of exhaustion. Now watching my own children raise their children, I see how important being involved with a community is and having supportive friends. I think we were working so hard just to live. We were isolated and the community was small making it difficult to find and make close friends. I needed community and friends.

We learned many skills that we still use. We continued to raise our kids frugally. We have rehabbed four houses, and we love working together on projects. The farm helped us develop a strong relationship and a close family. It was a great adventure. We would do it again.

Epilogue

We finally sold the farm in 1983. All three of our children were in school, starting to play sports, and we were more involved with our jobs and community. It was time. We stayed in touch with the Lauritzen's, went to Dorothy's funeral, attended Larry, their youngest son's wedding, and then sadly his funeral two years later when his was killed in a freak farm accident.

Fast forward to 2020. We are living in uncertain times because of the COVID-19 pandemic. We are retired and in our early seventies, spending winters in Florida. In May, Roddy died unexpectedly in his sleep. It was heart related and so difficult for us all to comprehend and accept.

Because of the pandemic, my children are able to work virtually, and we have been able to spend weeks together. I decided to do special trips with my four older grandchildren. My grandson Liam, who is thirteen, and I decided to do a family history trip. We planned to drive to Faribault, Minnesota, where Rodney Mott lived in the 1850's. Roddy had been named after him, and all I really knew about Rodney Mott was that he was the father-in-law of Willis West, Roddy's great grandfather, and had started the school in Fairbault for the Deaf. We have since learned he was the mayor, a judge, superintendent of public schools, and ran the school for the deaf for forty years. He had walked to California during the Gold Rush, returned by way of Panama. He was the editor of an abolitionist and temperance newspaper, and his home was part of the underground railroad.

As we started driving from Hayward, I asked Liam if he would like to see our farm where his mother had been born. It is an hour south of

Hayward, so it was on the way. We hadn't been able to drive onto the property since we had sold it because there was always a huge gate across the driveway that was locked.

I said, "Okay, Liam, if the gate is open, we are going up to the house."

We all knew that the people who own it now had built a new large house which we always called a McMansion. Before we had always stared up at it, never being able to get close.

Today, the gate was open! The old red barn was gone, the pastures were beautiful, and the fields were filled with hay and corn. As Liam and I approached the house, we saw two cars parked inside the garage, so we knew someone was home. We put on our masks and climbed the porch steps and rang the bell. Nothing. I waited a minute and rang it again. Finally, a woman with two large dogs opened the door a few inches. I told her who I was, talking as fast as I could to give her as much of my life story so that she wouldn't slam the door. I told her we had lived there in the early seventies, and kept it until the eighties, that my husband had just died, and I was taking my grandson on a journey and that we were starting with where his mother was born. She looked at us cautiously, seemed a bit suspicious, and kept the door at a crack. Finally, she opened the door and stepped on the porch, shut the door, and left the dogs inside.

Then she began to tell us her story. They had bought the property in 2003. They had lived near Turtle Lake, an hour south and their land and house had been destroyed by a tornado. Her husband couldn't stand to live there anymore and see such devastation. She said she would move if he found a place with a lake, woods, and pastures. She said she couldn't believe he had found this beautiful place. They built a new house and developed wonderful gardens. It was in front of where our house had been. They wanted to see the lake. In thirty years, the trees had filled in. They had abandoned our house and tore it down because her grandchildren liked to play in it, break windows, and she was afraid they might get hurt.

I felt uncomfortable being there. I was filled with so many emotions and so many memories. My body was shaking. It is such beautiful land, and these new owners are taking such good care of it. It made me happy. At the same time, I could picture Roddy and my children there with me. I wanted to walk to all of our favorite spots, but I felt like an intruder and was nervous that her husband would be startled by us if he saw us walking around.

She told me her name was Pam, no last name. I would love to go back again, but I don't think it will happen. Maybe it is best to hold onto the memories of our farm adventure and be content. We had lived the good life.

Recipes: Bread And Butter Pickles

When we lived on the farm, I loved making pickles, especially picking and sorting the cucumbers. I used the small ones for little dills and the medium ones for sweet-and-sour pickles. Thinly slicing the vegetables was a huge process for the bread-and-butter pickles. All of the vegetables were covered with salt and left to stand on the counter overnight. The smell permeated the air. They were perfect gifts to give.

25-30 medium cucumbers
8 large white onions
2 large sweet peppers
½ cup pickling salt

5 cups cider vinegar
5 cups sugar
2 tbsp. mustard seed
1 tsp. turmeric
1 tsp. celery seed
½ tsp. cloves

Wash cucumbers and slice thinly. Chop onions and peppers. Combine these and cover with salt and let stand overnight. Drain. Combine cider vinegar, sugar and spices in a pan and boil. Pour over drained cucumbers in jars. Use sterilized lids and rings. Place the jars in the canning kettle covered with water. Bring to a boil. Boil for ten to twelve minutes.

Recipes: Top Me Twice Cake

The Top Me Twice Cake was a favorite because it was customary to have a pan of bars, cookies, or a cake on hand when people stopped by for coffee. This was a custom I had never known about. In my family you always called ahead before you arrived at someone's house; and you were coming for a purpose, not just to visit. Roddy had quite a sweet tooth, so we could go through a pan of bars or dozens of cookies on a daily basis.

Cake:
2 cups flour
1 cup sugar
1 tsp. soda
1 tsp. salt
1 can (13 ½ oz.) undrained pineapple
1 tsp. vanilla
2 eggs
Topping:
½ cup firmly packed brown sugar
½ cup flaked coconut
½ cup chopped nuts
Sauce:
½ cup butter
½ cup light cream
½ cup sugar
½ tsp. vanilla

Top Me Twice Cake (Continued)

Preheat oven to 350 degrees. Grease bottom only of a nine-inch square pan.

In a large bowl, sift dry ingredients together and then combine with remaining cake ingredients at lowest speed of mixer until blended. Then beat for 2 minutes.

Pour batter into a nine-inch square pan.

Combine topping ingredients and sprinkle on cake batter.

Bake for 45 to 50 minutes.

Just before the cake is done, prepare the sauce. Melt the butter in a small saucepan. Blend in the remaining sauce ingredients.
Poke holes in the cake and pour sauce over the warm cake. Cool before serving.

Other True Stories

Meeting Mr. Wright

My father was finishing his fellowship in internal medicine at the Mayo Clinic, and he was deciding where he wanted to practice medicine. Our family plus a babysitter, Mrs. Scroggins, took the train to Phoenix. I was three and had two older sisters eight and ten. The babysitter was an older woman from rural Minnesota who took a dislike to me. Every day she would brush my hair, yanking and pulling it. I would cry and complain. When I took my nap, she refused to give me my bottle, saying I was not a baby and didn't need a bottle.

My parents had left early for an important interview. Mrs. Scroggins grabbed the hairbrush and started to try and tame my wild hair. I was squirming and screaming, "Ouch! You're hurting me!"

My older sister, Sheila, stepped right up, inches from Mrs. Scroggins's face and said, "If you're going to be mean to Wendy, you can leave right now."

Mrs. Scroggins walked into the other room, grabbed her suitcase, and walked out the door, never to be seen again.

This created somewhat of a problem for my parents. They had many meetings scheduled, so they decided to take me along when they were going to Taliesin West to meet Frank Lloyd Wright. I had a satchel of toys and was told by my mother I was to play quietly and not disturb the meeting. I was happy to be with my parents and played. After about an hour, Mr. Wright said to me, "Wendy, what do you think of my house?"

I looked up at him and clearly said, "I think it's crooked."

My parents did not choose him as their architect and always loved telling the story. I wish I could remember what Mr. Wright and his house looked like.

I Never Told Anyone

When I was six, my parents were building a house in the country. It was going to be a big, beautiful, very modern house made of concrete, with chrome columns, travertine floors, and hardware made of stainless steel. My father had picked the land by climbing trees to get the best view of the valley. He bargained with three farmers and bought parcels of their woods. It was going to be a wonderland for our family of five children.

Every week we would go out to see the progress and bring a picnic. The girls' bedrooms were going to be in the lower level, and there would be a huge playroom with a walkout. We were promised that we would get a television. My sisters were able to pick the color of their walls and bedspreads, but my mother picked mine. The walls were so pale pink you couldn't tell they were pink. My bedspread was white quilted chintz with a pale pink dust ruffle. I hated it.

One Sunday I was following behind my father as he walked on planks above the lower level, checking dimensions and where the fireplace would go. There was a box of two-inch nails sitting on a table saw, and I picked one up and put it in my mouth. It was cold and had a metallic taste. All of a sudden my father turned around to say something to me. I panicked and swallowed the nail. I was afraid he would see it in my mouth and get really angry. His specialty was endoscopy, and he regularly showed us things like pennies, peanuts, and pieces of field corn that he had retrieved from a patient's bronchial tube. He was very strict about us running with things in our mouths or eating things we could choke on.

I didn't tell anyone for years, until I was eleven. Sometimes I worried that the nail was stuck in my lung or stomach or that it would perforate my intestines. Finally, when I did tell my father, he said, "You could have just checked your feces for it." Oh, how gross, I thought. I would be so embarrassed if anyone knew I was checking my feces.

Painting the Porch

My best friend Joanie was at my house. She lived down the block and was two years older than me. There were eight children in her family, so something fun was always happening.

Our yardman, Mr. Phillips, had painted our front porch grey about a week before. It had been a pretty brick red, but my mother thought it showed too much dirt and dust.

"I really don't like this color," I said to Joanie as we sat, bored, on our porch.

"I think we should see if there is any paint in the garage that we could use to make it look a little brighter."

We found some of the same brick red. I said, "We can paint it! Let's go."

We started painting and had covered quite a bit of the open side of the porch. We had dribbled some drips on the wall and quite a bit on our clothes. I thought maybe this wasn't such a good idea. We quickly put the paint back in the garage and threw the brushes into the trash. We stayed outside and played.

All of a sudden my dad's car pulled into the driveway, and he started to walk up to the house. He stopped at the steps. He looked at the porch. Joanie and I were watching from underneath the neighbor's pine tree. We were really scared, and we both started whimpering.

"Wendy, come here. I want to talk to you," he said in a very stern voice. We slowly walked up, and I said, "I'm sorry. I'll never do it again."

He sent Joanie home and told me to go to my room. He didn't want to know why. When he came inside, he called Mr. Phillips to come and repaint the porch.

Eavesdropping

When I was seven years old, my mother took me to New York for my birthday. I was so excited to be going to New York to spend time just with my mom.

There was a mix-up at our hotel. They didn't hold our reservation because we arrived late. It made my mother furious.

"Come on, Wendy. We're going to the St. Regis."

That actually meant nothing to me, but when we got there I was in awe. Our room had the most beautiful flowered chintz bedspreads I had ever seen, and there were huge white fluffy towels in the bathroom.

"Let's get room service. It's too late to go out," my mother said, reaching for the phone beside the bed. I looked over the menu and decided on a hamburger and French fries. I was a bit nervous because it cost $12.00.

My mother said, "Don't worry. It's a special occasion." Room service arrived and a man pushed a little white table with fresh flowers and the plates covered with silver domes to keep everything warm.

In the morning we went to FAO Schwarz. It was three floors of toys, games, and books. There were huge dark wooden tables that had all sorts of games and toys on them for people to try.

My mother walked me around the whole store and said, "I have an errand to run. I want you to stay here and play until I get back. Then you can pick out your favorite toy, and we'll have it sent home to Minnesota. Do not leave the store, just investigate everything."

I started with the dolls. There were so many and they had such beautiful clothes. Next I went to the board game area, but I wasn't that interested because there was no one to play with. The other customers and clerks didn't seem to notice me at all. Then I came to a section I'll call dramatic or creative play. There were small shops, dollhouses, and stand-alone kitchens. I loved all of the miniature food, pots and pans, and the idea of my own little shop.

All of a sudden, my mother appeared. I hadn't realized she had been gone for a long time. We bought the small shop with miniature food and utensils. I couldn't wait to play with it at home.

We did some shopping at Bonwit Teller's, where my mom bought me a beautiful navy-blue cardigan.

"Navy is an elegant color, Wendy, and it will look so lovely with your dark hair and blue eyes."

We went to Brentano's bookstore and my mother bought the French magazine *Paris Match* and *The Secret Garden* for me.

"Now let's go to the St. Regis and have some tea," she said.

We walked into the dark lobby bar area with stuffed brown leather chairs and small round tables.

"What I love to do," my mother said, "is to listen to other people's conversations. It's kind of fun to hear what they are talking about and imagine what their lives are like."

A man and a woman came in and sat at the table across from us. I could clearly see both of them. They ordered drinks.

"Those are martinis," my mother said.

I sat very quietly and waited for them to say something. I was pretending that I didn't see them at all.

"Will you leave him? I want to be with you forever," he said quietly.

She said, "I want to be with you more than anything. I hate my husband, but how can I leave him? What will happen about my kids? If he finds out about us, I know he'll take the kids from me."

I was really getting confused. "What do you think is going on, Mom?"

"Well, Wendy, they are having an affair."

"What does that mean? Why would her husband take her kids?"

"It's complicated," my mother said as she finished her tea. "Let's go freshen up and get ready to go to the play and out to dinner."

That was the beginning of our lifelong habit of listening to the next table.

The Russians Are Coming

We're playing kickball and the bell rings. Everyone is sweaty and smells of thawing dirt. We file in, laughing about the mud on our uniforms, and quickly form a single line, walking with our arms folded. We sit quietly in our desks. The room smells of chalk, dust, and heat.

Sister Toma is tall, skinny, and as white as the white in her habit. She is standing in the front of the room staring at us.

"You're all going to die! You'll have to die for your faith. You have to tell them you believe in God. Tell them you are Catholic. The Russians don't believe in God, and they are taking over our country."

Sister then starts our math lesson and nothing more about the Russians is mentioned. I forget about the Russians at Grandma's house after school.

"Wendy, there is coffee cake in the ice box and some Red Delicious apples. Have as much as you want, dearie. You can watch TV until your dad comes."

As I am getting ready for bed, I remember what Sister said. I wonder do I have to tell them? How will they know? Is there a way you can tell by looking at me?

"Wendy, let's read *Little Women* now," my mom says as she opens my bedroom door.

I burst into tears and say, "Do I have to tell them?"

"Who? Tell them what?"

"The Russians. That I believe in God. They are going to kill me."

"Wendy, what are you talking about? "

"Sister says we have to die for our faith. They're coming and will ask us if we are Catholic."

"God no, that is ridiculous. They aren't coming here. They just put Sputnik into space and will be concentrating on that."

I am so relieved. I fall asleep after my mother reads the first page.

Violet

My sisters both had braces. Actually, they had finished wearing braces and now had retainers which we called turtles. They would push their tongues on the roof of their mouths and make this incredibly deep clucking sound. I wanted to be able to do that.

Dr. Wentworth, a wiry, balding man with a big nose and beady eyes, was the orthodontist who came to Rochester every two weeks on Monday and Tuesday and fitted and adjusted everyone in town who needed braces. One Monday morning in early June, I went to his office with my sisters. I signed in just as they did and took my seat in the stuffy, crowded waiting room.

Finally the receptionist called my name, and I was shuttled back to an examining room. Dr. Wentworth looked at my teeth. Then he had me open wide and bite down so he could check my bite. With his thumb and index finger, he popped my eyeteeth out. They were baby teeth, not loose, and it didn't hurt much. He fitted each one of my front teeth, uppers and lowers, with white gold bands. He told me to come back in two weeks. My teeth were sore from all of the poking and fitting, but I suddenly felt older and more sophisticated.

My sisters and I went to the pool for the afternoon and arrived home about five. I later learned that sometime during the afternoon, Dr. Wentworth called my mother and said, "Violet was here this morning, and I have fit her for braces."

My mother said, "I'm sorry, I don't know who you are talking about. I don't have a daughter named Violet."

"Well, Violet came in here this morning with Sheila and Maureen and signed in as Violet Donoghue. They acted as if she was their sister."

"My youngest daughter is named Wendy. I will find out what is going on when she gets home," my mother said and hung up the phone.

When I walked through the kitchen, my mother said, "Wendy, do you know anyone named Violet?"

I smiled a huge grin, showing off all of my braces.

"I don't think your father is going to be very happy about this. You had better think of a good explanation before he gets home."

All I could think about was how old I now looked and how in eighteen months I would have a turtle and be able to make that incredible clicking sound.

My First Job

When I had just turned twelve and was going into seventh grade, I met a new friend, Cindy. We decided that we would start a catering business. We would cater parties for children. We would do everything: plan the food, the games, and the prizes. I had a small black three-ring notebook that I used to show our clients. I typed up party themes, lists of games, food options, and a variety of prizes people could choose. We ordered cakes from Arbuckle's, the best bakery in town.

We went around and knocked on doors in Cindy's neighborhood, Pill Hill. We only went to houses where we knew the kids or the parents were friends of our parents. On the first day we got two parties. The first was a birthday party for Dodo Ellis. Her real name was Laura. She was one of eight and that's what her family called her. She was turning six.

It was going to be a simple party with cake and ice cream. There were going to be twelve children. Half of them were her siblings. We ordered the cake, and it was decorated with lots of flowers, and we served Neapolitan ice cream. The paper plates, napkins, and cups that we picked out at Woolworth's were decorated with pink and white polka dots.

We played pin the tail on the donkey and musical chairs and had free play in the front yard. We worked really hard; everything went smoothly, and it was a huge success.

Our second party was a neighborhood party, "Farewell to Freedom," to be held at the Blackburn's, who were Cindy's neighbors. The theme was Mrs. Blackburn's idea because all of the kids were going back to school. She was really excited about our ideas.

We made tuna fish salad that we put in hot dog buns and put little flags on toothpicks so that they looked like boats. We served potato chips and grape juice. We ordered a sheet cake that said "Farewell to Freedom," decorated with children walking toward a school. The games were all outside. We had a relay where the kids had to put on oversized clothes and hop down to the other end of the yard, then take off the clothes and have another team member put them on. We played pin the flag on the school. We also had a game where the kids walked around in a circle; and when the music stopped, they stood on a number, and we would call out the winning number.

The prizes were from Woolworths, things like bubbles, sets of jacks, packs of cards, and my favorite, surprise balls. They are balls made of layers of crepe paper that have little prizes in the layers. The center had a bigger prize like a rubber ball or a mind-bending puzzle. There were probably twenty kids at the party, and everybody loved it and had a great time.

The third and last party was at my house for my twin brothers. They were seven. We had cupcakes and lemonade and peppermint candy ice cream. The first game was a relay race where the kids had to walk from one end of the yard with a serving spoon with three cotton balls stuck together. They couldn't drop the cotton balls and had to pass them to the next person. We had wheelbarrow races, and for the last game there were squirt guns with which they had to try and shoot out a lit candle. The guns were the prizes they took home.

It was a disaster. My brothers wouldn't listen when I tried to explain what to do.

"We don't want to play that game. You're so bossy," Edmund said.

I grabbed him by the arm and pinched him to make him do what I wanted, and he started screaming and crying and yelling for my mom. The other boys just started running around. The whole thing fell apart.

Cindy and I were exhausted and were so happy when it was time for everyone to go home. We never really talked about doing another party. I wish I still had my black notebook.

PRANK CALLS

When I was in the sixth grade I met a new friend, Kaia. Kaia lived in town in the area known as Pill Hill, a neighborhood with tons of kids where all of the doctors lived. We lived way out in the country and our nearest neighbors were a quarter of a mile away and there were no kids my age. I would spend the night at Kaia's on Fridays. She was the oldest in her family, had her own room, and we didn't see much of her brother and sister, or her parents either.

Our favorite thing to do was make prank calls. This was so exciting for me because my father had a rule that I could only talk on the phone for three minutes. He was on call a lot and said it would be very dangerous for me to tie up the phone in case there was an emergency. He would say I had telephone-itis and the cure would be to take the phone away.

We would take the phone into Kaia's room and close the door. We'd call random numbers using different voices by plugging our noses and saying, "Hello, this is the electric company. Is your refrigerator running?" Some people would say yes and we would say, "Well, you had better catch it!" We would laugh and laugh.

My favorite prank was to pick a name out of the phone book by closing my eyes and running my finger down the page and stopping. If the person answered, I would say, "Is Beth there?"

"I think you have the wrong number."

"Oh, Beth told me to call her at this number. Will you please tell her Shelly called? Thank you."

We would wait about half an hour and call back and say the same thing. Sometimes we would add a little information. "Beth told me to call at this number after I finished dinner. Will you please tell her Shelly called?"

We took turns so we each had several of these calls going on. After another half hour we would call back and say, "Hi, this is Beth. Have there been any calls for me?'

It was amazing the number of people who would say, "Yes, Shelly called. She has finished her dinner and can meet you now."

The Door

We lived in the country, seven miles from the clinic door on a ridge of maples, birch, and oak. We could see the farm fields below. My father bought the sixty-five acres from three different farmers at one hundred dollars an acre. He climbed trees to decide where to place our house.

It was a beautiful house and the grounds were beautiful too. We had trails for riding and walking through the woods. We could walk to the pasture by the road, to the apple trees and wild grape vines, to the upper forty along huge fields of hay, or down into the ravine filled with oaks and maples.

Our driveway was a quarter of a mile long. We had one Doberman, then another, and finally two Irish Wolfhounds, because my father thought they looked menacing if anyone drove down our driveway. The dogs would be alert, run out through the dog door, and stand in front of the house barking.

When I was old enough to stay alone, I remember my mother telling me: "If you ever feel afraid or feel there is someone on our property, just walk out the side door in the laundry room and go into the woods. You know your way around the woods, and you'll be safe."

I never did it, but the memory of her telling me and the feeling of walking in the woods always stayed with me.

One time two men drove down our driveway and knocked on the door. My father was home and answered the door. They wanted permission to hunt on our land. It was not possible because my dad had the property declared a state game refuge. He liked to raise pheasants and turkeys, band them, and release them.

He told the men they could walk through our property and that he would show them the way. He went back to his den and got a huge hunting knife, which he tucked into his sock. I watched him do it and asked, "Why are you taking a knife?"

"You can never be sure what people are up to, what their motives are." I watched them walk up behind the barn and wondered if I should go out the side door into the woods to be safe.

John's Call

It was August and summer vacation was coming to an end. My twin brothers were having a hard time finding things to keep them occupied. Edmund, the quiet one, liked to be outside. John was more gregarious and needed action. My mother left them alone to go and get her hair colored, cut, and styled. It was a process that took a few hours.

As my mother sat under the large domed hair dryer, the stylist came and said there was a phone call for her. She went up to the receptionist's desk and took the phone.

"Mom—at the end of the driveway—a truck—Edmund—in the go-cart," John cried in a frantic voice.

"What? What, John, what are you saying?" my mother said, beginning to panic.

We lived seven miles out in the country on a gravel road and our drive was a quarter of a mile long. The boys had a go–cart and would ride for hours up and down the driveway, but they were forbidden to go outside of the old farm gates to the gravel road.

"Edmund—at the end of the road," John repeated in a desperate voice.

"John, are you serious? Is Edmund hurt?" My mother sounded hysterical.

"Just kidding," John said and started laughing nervously.

"Wait until I get home, John," my mother said in a sharp tone. She arrived home about forty minutes later.

John was sent to his room, not allowed to watch TV or have friends over. This punishment lasted about a week, but the incident has been part of our family folklore for over fifty years.

I Paint a Picture in My Head

I paint a picture in my head. I'm tall with smooth, soft, straight hair that falls gently to my shoulders. It stays that way in the morning humidity as I walk to school.

My best friend has thick blond hair that falls on her shoulders. It always looks beautiful. It's not affected by the wind, humidity, or dryness of the air.

I read about girls who straighten their hair in *Glamour*. Some do it by using a stinky permanent solution that's combed for forty-five minutes. Others use stronger chemicals like lye. You must go to a beauty shop, preferably one that has African American clients. It costs eighty dollars, which is a lot of money for a thirteen-year-old girl who makes fifty cents an hour babysitting.

The cheapest and quickest way to straighten your hair, I learn, is by ironing it. You use a comb and section the hair into bundles about an inch wide. You set the iron on rayon. The place to start is at the roots. After a section is ironed, you place a rubber band to keep it from the unfinished sections.

I decide this is what I am going to try. Our iron is old and heavy, but I think the weight might be a plus. I start in the front. I do two sections. They look great, so smooth and straight. I'm about to do the section just above my left ear. The tip of the iron grazes my ear. I jerk my head up and just then the iron touches one of the rubber bands – sizzle, stink. I quickly try to get the melted rubber band unglued from my hair. I end up pulling out a huge clump of hair. A blister is forming on my ear.

I straighten a long section from the crown of my head. I lay the hair out on the ironing board. Ooh, what's that smell?

It's my hair singed and frizzled! The whole room reeks! I check the iron and realize the dial has slipped to cotton. I run to the kitchen to get ice for my ear.

My grandmother is in the kitchen. She looks up at me.

"Wendy, what have you done to your hair? Oh my, it looks awful! Why did you do that to your beautiful curly hair? All you need is a ribbon to pull it back from your face."

Mirror, Mirror on the Wall, Who is the Fairest of Them All?

My mother used to say to me, "Wendy, wear a big sun hat. Don't get your face in the sun. You have such beautiful white skin, dark hair, and blue eyes."

Of course I wanted to be tan. I mixed iodine with baby oil and would bake in the sun for hours. I thought my color was fine, not as tan as my brothers, but as dark as I could get.

When I was nineteen, I spent the summer as an exchange student in India. I lived with a family in Lucknow, in north central India. The family was warm and friendly. There was a sister, Rani, my age and a brother, Inder, two years younger. They had two house servants, Ayah, the maid, and Chunder, the houseboy. He was about seven, always smiled a lot, and watched me closely. Ayah, on the other hand, was very quiet and kept to herself. I loved her ankle bracelets, toe rings, and the beautiful thick silver bracelets she wore on both arms.

After I had been living in the house for several weeks, we had settled into a routine. In the afternoon we had tea and special sponge cakes at about four. Ayah would serve us. One day she looked straight at me, covered her mouth, and began laughing hysterically.

"What's so funny?" I asked. "Why is Ayah laughing?"

Rani and Inder were silent, their eyes cast down.

"What is it? Did I say something bad?" Several minutes passed and Rani said, "She thinks you look like a newborn pig. Your skin is so light and pink."

There was lots of nervous laughter. Ayah thought dark skin was more attractive than light skin.

What did my mother know?

Grandma Tells About Her Babysitting Adventure

Sunday afternoon I had a call from my daughter, Wendy, that their family had spent the night in the emergency room with Brendhan. They thought he had the croup and took him into a steam-filled bathroom, and he was choking for air. The ER doctor said he had asthma, gave him an injection to open his airways, and said it would affect him for the rest of his life. Brendhan was twenty months old, very active, and not very verbal.

I agreed to drive to Minneapolis each day and watch him while Wendy worked. It would be an adventure. I hadn't spent much time with just Bebo, which was his nickname.

I arrived at 7:30, and Bebo was excited to show me his trucks that he called go-go's. He wasn't the least bit fazed when Wendy walked out the door with the girls.

I had left some of my books and a treat I had brought for Brendhan in the car.

"B, wait right here while Grandma goes to her car to get some special things."

Brendhan followed me out the front door onto the porch. As soon as I was down the three steps, he slammed the porch door shut and it locked.

"Open the door for me, please," I said with my arms full.

Brendhan looked at me through the glass and said, "House."

I tried to talk to him through the mail slot, “Please open the door.”

What was I going to do? The wood stove was burning inside.

What could he get into? Wendy probably wasn’t even at work yet. What would she think of her fine babysitter mother?

I kept trying to talk to him through the mail slot and he kept saying, “House. House.”

I ran to the neighbor’s, worrying about leaving him. Margot didn’t have a key, but said she would call Wendy at work and let her know. I ran back to the house and couldn’t see Brendhan. I was panicking, screaming his name into the mail slot. He came walking slowly, carrying a box of graham crackers.

“Oh, let’s have a party,” I said. “Open the door and we’ll have a party.”

He managed to get a cracker out of the package and put pieces of it through the mail slot.

Oh well, I resigned myself, I’ll just sit here freezing, eating graham crackers until Wendy gets home and can unlock the door.

It's Quitting Time

"No 'moke, Mama!" When Brendhan said that to me when he was two and barely speaking, I decided I needed to do something.

I found a hypnotist named Roger through a good friend who had successfully quit smoking. I called and took the first appointment I could get. It was a week away and scheduled for 3 p.m. I chain-smoked every cigarette I had before I arrived at his office.

I walked through the door and no one was there. A few minutes later a tall, thin man carrying a toothbrush and toothpaste walked in and said, "Can I help you?"

"I'm here to be hypnotized to quit smoking."

He had a puzzled look on his face and said, "I have you down for next week."

I almost burst into tears. He realized I was very stressed and anxious and said, "Well, let's see if my next client is coming. He's a bit late, so I'll call him."

Whoever answered the phone wouldn't acknowledge there was anyone by that name available. Roger said, "Well, I guess you're in luck. We can get started."

I sat in a puffy Lazy Boy chair and he covered me with a soft fleece blanket. He asked me how I used cigarettes.

"They're rewards," I said. "If I fold the laundry, I can have a cigarette. If I finish cleaning, I can have a cigarette. If I put the kids to bed, I can have a cigarette, and so on and so on. "

He then asked me about my work. At the time, I was teaching parenting classes and doing home visits with parents of infants.

Roger said, "I want you to think of a place you love to be. Visualize what it looks like, what it smells like, how you feel there. Now I want you to think about infants in the hospital being born. The birth process is hard work. Getting through the birth canal takes a lot of energy. Now, once the infants are born, I want you to give them a reward for their hard work. Put them in their bassinets and give each newborn a cigarette. Light it and put it in their mouths."

I was to alternate this visualization with the visualization of the place I loved to be, back and forth, over and over. Two hours had passed, and Roger said, "We're finished for today." He handed me a recording that he had made of my visualizations and said, "Continue to listen to these. It will take time. Don't get discouraged. We can meet whenever you feel the need to."

I left the office, went home, and thought about my visualizations. If I had the urge to smoke, I thought of the infants, how they looked with cigarettes in their mouths, and smoke all around them. I never listened to the tape. I never went back for another session, and I have never smoked another cigarette. Roger never charged me.

Far, Far Away

An older boy asked Eric, a three-year-old in Brendhan's preschool, if he would like a ride on his brand-new ten-speed bike. The older boy was a neighbor, and Eric said yes. That boy took Eric to a garage where he molested him.

The teachers at the preschool notified all of the parents about the incident. I had a conversation with Brendhan about not going with people unless he made sure his father or I knew he was going somewhere. I taught him the rhyme, "Say no. Don't go. Run away and tell." We talked about it quite a bit, and I explained what had happened to Eric.

About a week later I was cleaning on my front porch, and Brendhan was outside riding his Big Wheel up and down the sidewalk. An elderly woman walked by Brendhan and said, "Hello." He looked up at her from his Big Wheel and said, "Don't you know better than to talk to a stranger? I might take you far, far away and pull down your pants."

She had a stunned look on her face and looked up at our house. I crouched down to the floor laughing. She walked briskly past Brendhan, who just kept riding his Big Wheel up and down the sidewalk.

Standing on the Edge

Brendhan was having a birthday party. He was turning twelve and had invited eight boys to spend the night. We reserved movies, ordered pizza, and I made a cake decorated with hockey goals and players.

The boys went into the basement to watch their movies. I was in the kitchen cleaning up, and I heard this click, click, and click. We had an open area where I could see into the large room where the boys were. That's when I saw the gun. Damon was holding it and pulling the trigger. Feelings of panic, fear, and anger rose up in me. I walked slowly down the basement stairs; and when I reached the room, I said, "Damon, where did you get that gun?"

"My brother gave it to me. It's just a BB gun."

"You can't have a gun like that, Damon," I said in a soft, pleading tone.

"If anyone were to see you with that gun, like the police, they might shoot first and ask questions later."

The thought of a twelve-year-old walking around with a gun gave me the chills.

"Please give me the gun. You can't have it, and I don't want it in my house."

All of the boys stared, some gazing at the floor, looking anywhere except at Damon or me. He handed me the gun. As I carried it upstairs, I was shaking. It was a BB gun, but it looked like a 9mm pistol. I hid it in my closet between some sweaters.

Brendhan had known Damon in sixth grade, and they were now at the same junior high. He wasn't the kind of kid Brendhan usually hung out with. Brendhan had gotten to know Damon through Brendhan's role as a peer mediator. Brendhan had been chosen to be a peer mediator by the assistant principal. If an altercation broke out between students, he would be called to interview the students separately and help to create a solution to solve the problem. If things were not resolved, the assistant principal would take over. Brendhan enjoyed the training and being called to help settle fights and arguments. It gave him a chance to interact with rougher kids, something he was extremely curious about. He was a popular boy, and peer mediation allowed him to mix with all kinds of kids in a safe way. It also got him out of class.

Damon was really having some trouble. His older brother, Ricky, had joined a gang. His mother was often strung out on drugs. Several days after the party, Damon came to our house after school.

"Can I stay at your house? My mom's been hittin' and whippin' me and I don't wanna go home."

"I'll need to call your mother so she knows where you are," I said. I felt sorry for him. I dialed his number and a woman answered.

"Hi, this is Wendy West, Brendhan's mother. Damon is here, and I was wondering if it was all right with you if he spends the night."

"You tell that boy to get his ass home now. My son is not staying at your place," she growled.

I drove Damon home, wondering what would happen to him. The next morning, I went to the school to talk with the social worker.

That afternoon, when I got home from work, someone had been in my bedroom. The gun was gone as well as a pair of Brendhan's jeans.

Damon disappeared. He wasn't in school the next day or ever again.

Shoes

I love shoes. When I was working and if I had a particularly rough day—lots of drama or trauma—I would get in my car and go to DSW, Designer Shoe Warehouse. There are rows and rows of shoes, starting with athletic shoes, then boots, sandals, sensible shoes, shoes for travel, and then high heels and really high heels.

I start at the athletic shoes, light, bright, perfect for running and cross-training. You feel fit just looking at them. I usually don't buy these because I came for something frivolous, without a purpose, to change my mood. If it has been a really stressful day, I may buy two pairs.

I walk past the boots, no need for them. I hope to be spending more time in Florida. Next come the sandals. I always like these, especially ones with thin straps that float on your feet. I don't like things between my toes that slip on and are the flip-floppy type. I try on several pairs and carry one to the next section—sensible shoes. These are for traveling. I let my mind wander. What city will I be walking miles in? They need to be good for all day, but I will probably switch to sandals for the evening. Will they step on historic places? I have a pair of hiking boots I wore to walk the Great Wall. Whenever I see them in my closet I think of the people who built the wall, what their lives were like. Then I think of the young couple we met on the wall. They were graduate students in medicine, one a cardiologist and the other working on a cure for schizophrenia. They walked with us for four hours to learn about us, practice their English, and find out what we knew about graduate programs.

Next it's high heels. I like to look at these, but never buy them. I can't walk in high heels. My mother told me when I was a young teen, "If you wear heels you need to stand up straight, put your knees together, and keep your ankles stiff. Otherwise you look ridiculous."

That's what I think every time I try them on: "You look ridiculous."

I just buy the sandals and bring them home in the black and tan shopping bag. Roddy looks up from reading the paper and says, "Was it a one-pair or two-pair day?"

I tell him about my student who is being evicted from her apartment with her three children because she has allowed her husband, who just got out of jail and traveled from Illinois to be with his family, to move in. It violates the rules of low-income housing. I have found her a lawyer, and we will be going to court in ten days. Her children can't understand why their daddy can't stay and I am having trouble finding him a safe place to stay.

Now I have worked myself up. Should I go back and find some sensible traveling shoes?

Acknowledgments

I wish to thank my mother, Winnie Donoghue, for my wonderful childhood, and my dear friend, Vicky Lettmann, for encouraging me to love writing and to finish this project.

I also want to thank the members of my memoir writing group and my classmates from the Joy of Writing class at BIG Arts on Sanibel Island, Florida. A thank you to Arlene MacDonald and also to Bryan Williams and Chad Cross at Panther Printing.

Author's Note

I started writing family stories in 2004 when I took a memoir class. I continued in Vicky's Joy of Writing class where we used the prompt method of writing and would write for ten minutes. If we liked what we had written, we could expand on it. If not, it was just a ten-minute exercise.

I began writing stories for my grandchildren. In the beginning, I would write about experiences I had when I was their age. I have always loved stories and love being a storyteller. I hope my family and friends will enjoy reading these stories as much as I have enjoyed writing them.

About the Author

Wendy West is a native Minnesotan, who now spends half her time in Florida where she writes stories for her children and grandchildren. She and her late husband Roddy have biked through Vietnam, China, India, and Madagascar. In 2018, they hiked in Patagonia. Their last biking trip together was to Spain in March, 2020. Wendy is an avid rug hooker, who is always looking for adventure.